The publishing house tradition has created the series **TREDITION CLASSICS**. It contains classical literature works from over two thousand years. Most of these titles have been out of print and off the bookstore shelves for decades.

The book series is intended to preserve the cultural legacy and to promote the timeless works of classical literature. As a reader of a **TREDITION CLASSICS** book, the reader supports the mission to save many of the amazing works of world literature from oblivion.

The symbol of **TREDITION CLASSICS** is Johannes Gutenberg (1400 – 1468), the inventor of movable type printing.

With the series, tradition intends to make thousands of international literature classics available in printed format again – worldwide.

All books are available at book retailers worldwide in paperback and in hardcover. For more information please visit: www.tredition.com

tradition was established in 2006 by Sandra Latusseck and Soenke Schulz. Based in Hamburg, Germany, tradition offers publishing solutions to authors and publishing houses, combined with worldwide distribution of printed and digital book content. tradition is uniquely positioned to enable authors and publishing houses to create books on their own terms and without conventional manufacturing risks.

For more information please visit: www.tredition.com

Bromide Printing and Enlarging A Practical Guide to the Making of Bromide Prints by Contact and Bromide Enlarging by Daylight and Artificial Light, With the Toning of Bromide Prints and Enlargements

John A. Tennant

Imprint

This book is part of the TREDITION CLASSICS series.

Author: John A. Tennant
Cover design: toepferschumann, Berlin (Germany)

Publisher: tredition GmbH, Hamburg (Germany)
ISBN: 978-3-8491-4798-3

www.tredition.com
www.tredition.de

Copyright:
The content of this book is sourced from the public domain.

The intention of the TREDITION CLASSICS series is to make world literature in the public domain available in printed format. Literary enthusiasts and organizations worldwide have scanned and digitally edited the original texts. tredition has subsequently formatted and redesigned the content into a modern reading layout. Therefore, we cannot guarantee the exact reproduction of the original format of a particular historic edition. Please also note that no modifications have been made to the spelling, therefore it may differ from the orthography used today.

CONTENTS

- Chapter I
  VARIETIES OF BROMIDE PAPERS AND HOW TO CHOOSE AMONG THEM
- Chapter II
  THE QUESTION OF LIGHT AND ILLUMINATION
- Chapter III
  MAKING CONTACT PRINTS ON BROMIDE PAPER; PAPER NEGATIVES
- Chapter IV
  ENLARGING BY DAYLIGHT METHODS
- Chapter V
  ENLARGING BY ARTIFICIAL LIGHT
- Chapter VI
  DODGING, VIGNETTING, COMPOSITE PRINTING AND THE USE OF BOLTING SILK
- Chapter VII
  THE REDUCTION AND TONING OF BROMIDE PRINTS AND ENLARGEMENTS

Chapter I
VARIETIES OF BROMIDE PAPERS AND HOW TO CHOOSE AMONG THEM

Contents

What is bromide paper? It is simply paper coated with gelatino-bromide of silver emulsion, similar to that which, when coated on glass or other transparent support, forms the familiar dry-plate or film used in negative-making. The emulsion used in making bromide paper, however, is less rapid (less sensitive) than that used in the manufacture of plates or films of ordinary rapidity; hence bromide paper may be manipulated with more abundant light than would be safe with plates. It is used for making prints by contact with a negative in the ordinary printing frame, and as the simplest means for obtaining enlarged prints from small negatives. Sometimes bromide paper is spoken of as a development paper, because the picture-image does not print out during exposure, but requires to be developed, as in negative-making. The preparation of the paper is beyond the skill and equipment of the average photographer, but it may be readily obtained from dealers in photographic supplies.

What are the practical advantages of bromide paper? In the first place, it renders the photographer independent of daylight and weather as far as making prints is concerned. It has excellent "keeping" qualities, *i.e.*, it does not spoil or deteriorate as readily as other printing papers, even when stored without special care or precaution. Its manipulation is extremely simple, and closely resembles the development of a negative. It does not require a special sort of negative, but is adapted to give good prints from negatives widely different in quality. It is obtainable in any desired size, and with a great variety of surfaces, from extreme gloss to that of rough drawing paper. It offers great latitude in exposure and development, and yields, even in the hands of the novice, a greater percentage of good prints than any other printing paper in the market. It offers a range of tone from deepest black to the most delicate of platinotype grays, which may be modified to give a fair variety of color effects where this is desirable. It affords a simple means of making enlargements without the necessity of an enlarged negative. It gives us a ready means of producing many prints in a very short time, or, if desired, we may make a proof or enlargement from the negative fresh from the washing tray. And, finally, if we do our work faithfully and well, it will give us permanent prints.

The bromide papers available in this country at present are confined to those of the Eastman Kodak Company, the Defender Photo Supply Company and J. L. Lewis, the last handling English papers only. Better papers could not be desired. Broadly speaking, all bromide papers are made in a few well-defined varieties; in considering the manipulation of the papers made by a single firm, therefore, we practically cover all the papers in the market. As a matter of convenience, then, we will glance over the different varieties of bromide paper available, as represented by the Eastman papers, with the understanding that what is said of any one variety is generally applicable to papers of the same sort put out by other manufacturers.

First we have the *Standard* or ordinary bromide paper made for general use. This comes in five different weights: *A*, a thin paper with smooth surface, useful where detail is desirable; *B*, a heavier paper with smooth surface, for large prints or for illustration purposes; and *C*, a still heavier paper with a rough surface for broad effects and prints of large size. *BB*, heavy smooth double weight; *CC*, heavy, rough, double weight. Each of these varieties may be had in two grades, according to the negative in hand or the effect desired in the print, viz.: *hard*, for use with soft negatives where we desire to get vigor or contrast in the print, and *soft*, for use with hard negatives where softness of effect is desired in the print. For general use the *soft* grade is preferable, although it is advisable to have a supply of the *hard* paper at hand as useful in certain classes of work. The tones obtainable on the *Standard* paper range to pure black, and are acceptable for ordinary purposes. For pictorial work or special effects other papers are preferable.

Platino-Bromide paper gives delicate platinotype tones, and where negative, paper and manipulation are in harmony, the prints obtained on this paper will be indistinguishable from good platinotypes in quality and attractiveness. This paper comes in two weights, *Platino A*, a thin paper suitable for small prints, and having a smooth surface useful for detail-giving; and *Platino B*, a heavy paper with rough surface, peculiarly suited for large contact prints or enlargements. Both varieties are obtainable in *hard* or *soft* grades, characterized as above. *Matte Enamel*, medium weight; *Enameled*, medium weight; *Velvet*, medium weight.

Royal Bromide is a capital paper in its proper place, *i.e.*, for prints not smaller than 8 × 10 inches, and then only when breadth of effect is desired in the picture. It is a very heavy cream-colored paper, rough in texture, and giving black tones by development, but designed to give sepia or brown tones on a tinted ground by subsequent toning with a bath of hypo and alum. This paper, also, may be had in two grades for *hard* or *soft* effects; it is further adapted for being printed on through silk or bolting cloth, this modification adding to the effect of breadth ordinarily given by the paper itself. I have seen prints on this paper which were altogether pleasing, but subject and negative should be carefully considered in its use. Rough Buff papers are very similar in character. *Monox* Bromide, made by the Defender Photo Supply Company, is obtainable in six surfaces; No. 3, *Monox Rough*; No. 4, *Monox Gloss*; No. 5, *Monox Matte*; No. 6, *Monox Lustre*; No. 7, *Monox Buff*, heavy rough.

The Barnet bromide papers, comprising ten different varieties, differing in weight and surface texture but very similar to the kinds already described, are imported by J. L. Lewis, New York.

As a suggestion to the reader desiring to have at hand a stock of bromide papers, I would advise *Platino A*, or a similar *soft* paper for prints under 5 × 7 inches; *Matte-Enamel* for *soft* effects, or a similar paper, as an alternative; *Platino C* and *Royal Bromide* for *soft* effects, or similar papers, for prints 8 × 10 inches or larger, and for enlargements. To these might be added a package of *Standard B*, and another of one of the above varieties for *hard* effects, to complete a supply for general purposes. The beginner, however, will do well to avoid the indiscriminate use of several varieties of paper, although he is advised to get information of all the different varieties in the market. It is better to select that variety which is best suited to the general character of one's negatives and work, and to master that before changing to another. It is true that an expert can get more good prints on bromide paper, from a given number of bad negatives, than another expert can get with the same negatives and any other printing paper; but it is also true that for the best results on bromide paper the variety of paper used should be suitable for the negative. It will be found, however, that this word "suitable" covers, in bromide printing, a much wider range than is offered by many printing papers. In fact there are only two sorts of negatives which will not yield desirable prints on bromide paper: first, an exceedingly weak, thin negative lacking in contrast and altogether flat; and second, a very dense negative in which the contrasts are hopelessly emphatic. Even in such cases, however, it may be possible to modify the negatives and so get presentable prints.

The ideal negative for contact printing on bromide paper is one without excessive contrasts on the one hand, and without excessive flatness on the other. A moderately strong negative, such as will require from three to five minutes in the sunlight with a print out paper, fairly describes it. In other words, the negative should be fully exposed and so developed that there is a fair amount of density in the shadows. I have never been able, with bromide paper, to get the detail in the shadows of under-exposed negatives, such as we see in a good print made on glossy printout paper. For this reason the use of bromide papers with under-exposed negatives is not advisable. But there are a great many negatives which, while unsuitable as they come from the drying rack, can be easily adapted to the process by slight modifications. A very dense negative, for instance, may be reduced either with the ferricyanide of potash or persulphate of ammonia reducer; and a thin negative with proper graduations can frequently be intensified to advantage in the print. While, as has been said, there is great latitude in the matter of the negative, this latitude should only be availed of when necessary. Local reduction or intensification of the negative is seldom necessary, as better results can usually be obtained with bromide paper by dodging in the printing.

Chapter II
THE QUESTION OF LIGHT AND ILLUMINATION

Contents

Thus far we have gained a general understanding of the different papers and the characteristics desirable in negatives. Before we take up the actual manipulation of bromide paper there are a few elementary principles bearing on the important detail of illumination which we must master. These may necessitate a little thinking, but a practical grasp of them will make our after-work much easier, and ensure that fairly good prints from poor negatives will be the rule instead of the exception.

In the first place we have often read that a strong light overcomes contrasts, while a weak light increases them. Yet how many of us realize when we come to make prints by any process exactly what this means; in other words, how many of us apply the rule in everyday practice? It is very easy to see what is meant by the rule if we will take an ordinary negative, such as a landscape with clear sky, and hold it first six inches from a gas-flame and then six feet. It will be found in the first case that the sky portion is translucent while the clear glass will, of course, be clear; in the second the sky will be opaque and the clear glass still clear. The contrasts have been rendered greater by removing the negative further from the light-source. As this is true in the extreme case given, so it is true in a smaller degree where the distances are only slightly varied, as well as where we deal with the graded portions of the negative instead of with only clear glass and the densest portions. It is this fact that we utilize in bromide printing; and it is because we have such unlimited control over the strength of our light that it is possible with it to get equally good prints from a wide range of negatives. It is very much simpler and more practicable to regulate the strength of the light by increasing or diminishing its distance than by interposing sheets of paper, ground glass, or opal, as is occasionally done with other processes.

The necessity, however, for occasionally changing the strength of our light in this manner may seem to introduce an element of uncertainty into the problem of exposure; but there is another rule which brings it back again to simplicity itself, and enables us to quickly calculate equivalent exposures at varying distances from the light-source. This rule is: "The intensity of illumination varies inversely as the square of the distance from the source of light." For instance if a given negative requires five seconds exposure at one foot from the light, it will have an equivalent exposure if exposed for twenty seconds at two feet, the square of one being one, and of two being four.

It remains then only to apply these two rules to our actual work with bromide paper. The shadows in a certain negative will receive full exposure, say, in eight seconds at one foot from the light; but the high lights of the negative are so dense that no light will penetrate them at that distance from the light in that length of time. Hence a stronger light must be used, or the action of the same light continued for a longer time; but the latter will not do since the effect would be to over-expose the shadows. Hence, knowing that a strong light overcomes contrasts, we move the negative to the distance of six inches, where the rule tells us the equivalent exposure will be one-fourth that at twelve inches, in this case two seconds. Here the shadows get no more light, but it is possible that the high lights of the negative will be penetrated by reason of the additional force of the light.

On the other hand we have a thin, flat negative requiring for the shadows two seconds exposure at one foot from the light. Knowing that a weak light increases contrasts we move the negative three feet from the light, and instead of two, give eighteen seconds exposure, the rule telling us that this is equivalent. Thus we are enabled to regulate the strength of our light to suit the character of our negative. But a standard distance of one foot will not suit with all kinds of lights or with all sizes of negatives. If, for instance, our light is a Welsbach burner, giving an intense and comparatively white light, we will find that a normal negative will print too flat if exposed at one foot. In such a case two or even three feet would be a better standard. Experience with our light will, however, furnish the best standard, always taking a standard negative for the tests.

The size of the negative also has its influence on the unit of exposure. For instance, we may have a half-inch oil-burner, in which case we would probably have to expose a standard negative at four inches in order to get the proper contrasts. But this is out of the question with a negative of 5 × 7 or over, as a reference to the diagram, Fig. 1, will clearly show.

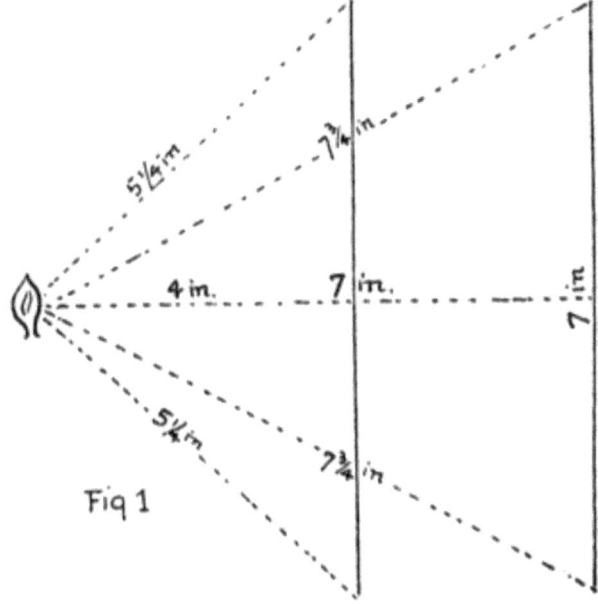

Fig 1

Here we find that while the centre of a negative is four inches from the light the extreme edges will be over five inches from it, the rule as to intensities telling us that the light at the edges will be only $^{16}/_{25}$ of that at the centre. This would result in a marked falling off of light at the corners, and would necessitate a constant motion of the printing frame throughout exposure, which is not wholly satisfactory. The remedy would be to use a stronger light at a greater distance. But another reference to Fig. 1 will show that if a 5 × 7 negative be held at seven inches from the light the difference will be only as 49 is to 56, which can in practice be disregarded, though it would be better to have it even less. Hence we see that it is never safe to have our unit less than the base-line of our plate, and it is better to have it even greater, as we will frequently be obliged to halve the distance to overcome contrasts. It follows from this that the larger our negatives the stronger must be our light.

Now all of these considerations may make very dry reading, but the reader who has followed them closely will see how vital they are to successful work. It should not be thought, however, that every exposure on bromide paper must involve an arithmetical calculation. On the contrary, once the proper distance from the light for the normal negative has been ascertained, it will be found that nine negatives out of ten will require no change in the distance from the source of light. This, of course, presuming that we classify our negatives and enlarge from those of the same quality at the same time.

One great objection to the use of bromide paper is that it must be handled in a dark-room. But this objection is not as serious as it may seem. An ordinary living room at night furnishes a delightful place in which to make prints, if we handle our solutions with reasonable care. The ruby glass can be removed from the dark-room lamp, and the orange glass used alone. But in this case, as indeed with the ruby light, care must be taken to guard against too much light. Development should be conducted at a distance of several feet from the light, and when almost completed, the tray can be brought close under the light to enable the worker to stop it at exactly the right moment. Ordinary bromide paper is about as sensitive as the process or slow dry plate or the average lantern-slide plate, and requires as much care as either, but not nearly so much as the most rapid dry plates. If fogging is noticed, of course additional precautions should be taken at once.

Chapter III
CONTACT PRINTING ON BROMIDE PAPER

Contents

Nothing more than will be found in an ordinary dark-room will be found necessary in bromide printing by contact, unless it be some arrangement for determining readily the distance of the negative from the source of light. For this purpose and with an oil-lamp, use a board a foot wide and about three feet long placed on the developing bench against the base of the dark-room lamp. It should be marked with black lines six inches apart. See Fig. 2.

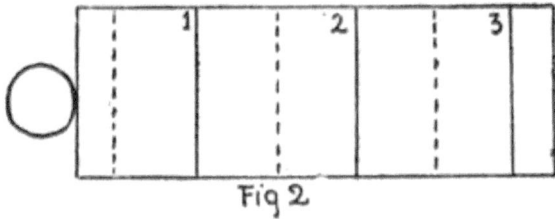

Fig 2.

Greater uniformity in lighting will be gained if a piece of white cardboard be placed immediately behind the flame. Some lamps have reflectors, in which case the card is unnecessary, provided that they reflect the light uniformly; otherwise such reflectors are worse than useless.

Having arranged the needful apparatus to our satisfaction, the last preparatory step before manipulation is the making up of a developer. Almost any of the modern developers (pyro excepted) will give good results with bromide paper. In every package of paper will be found the developers advised by the manufacturer of the paper used. Invariably there is among these a formula for ferrous oxalate developer. This is probably the best of all developers for pure black tones, but I cannot advise the novice to take it up in the early stages of his work with bromide paper.

When this developer is used an acid clearing bath is necessary, and this invites complications which may be disastrous to the prints. When experience has been gained, and a large number of prints are to be made at one time, it will be found advantageous as working longer with greater efficiency and more uniformity than some of the other developers. It is troublesome to prepare and does not keep well, apart from which there is the disadvantage that it does not permit of control in development in as large a measure as other developers.

A reliable metol and hydroquinone formula is as follows: Thoroughly dissolve metol, ¼ ounce; hydroquinone, ¼ ounce; in water, 80 ounces; add sulphite of soda (cryst.), 4 ounces; and carbonate of soda (cryst.), 2½ ounces. Bottled in 4-ounce vials and well corked, this developer retains its working power indefinitely. For normal exposures I take 2 ounces of the above and add to it 2 ounces of water. This will suffice for the development of three 8 × 10 sheets of paper, or their equivalent in smaller sheets. It is not wise to attempt to make it do more, as greenish tones will result. For the same reason, contrary to common opinion, I do not advise the addition of potassium bromide to the developer. It does not improve the developer, and may do harm.

An excellent developer which must be used freshly mixed, and may be made up in a moment, is as follows: Take 1½ ounces of a 25 per cent solution of sodium sulphite; dry amidol, 30 grains; 5 to 10 drops of a 10 per cent solution of potassium bromide, and dilute with 4½ ounces of water. A supply of new developer should be added as this is seen to become exhausted.

Other developing formulae could be given, but these two will be found to fill all requirements if properly compounded and intelligently used.

The greatest difficulty in developing bromide paper is to get rich black tones when desired, but this can be completely overcome by using entirely fresh developer from time to time, and never overworking the developer, whatever it may be. As compared with the paper, developer is cheap, and it is poor economy to save on the latter.

Except in rare instances the developer is better without any modifications whatever. In case of over-exposure, either general or partial, the developer after having been diluted as stated should be again diluted with its bulk of water. This gives blacker tones and more depth and life to the shadows. When through inadvertence we under-expose a print it may frequently be saved after partial development in the weak solution by flooding with a strong undiluted developer.

The temperature of the developer is of the greatest importance. In summer the aim should be to keep it approximately at 65 degrees Fahr., in winter, 70 degrees, but it should never be allowed to go over the latter. This can readily be accomplished by placing the graduate in a receptacle containing ice-water in summer or hot water in winter.

The paper is first opened at a safe distance from the dark-room light, and it is well at first to cut up one sheet into several slips to use as test-strips. If any difficulty is found in determining which is the sensitive side, it will be well to throw a piece of the paper on a plane surface when it will be seen that it has a slight tendency to curl. The concave is the sensitive side. Taking a standard negative we first take one of the test-slips and place it upon the negative so that it covers a portion containing both high lights and shadows. With an oil-lamp having a 1-inch burner, expose the test-strip behind the negative in the printing frame at one foot for ten seconds. Close the lamp and flood the exposed strip with the developer. The image should appear in a few seconds, and if properly exposed development will be completed in from one to two minutes, usually one. Rinse for a moment, and place the strip in a fixing bath made up by dissolving 3 ounces of hypo in 16 ounces of water. After a few moments examine the strip in full light, and see whether the contrasts are right. If so, expose a full sheet of paper, this time rinsing the exposed sheet before development to avoid the formation of air-bubbles. If the contrasts are too great try a strip at six inches from the light and two and a half seconds exposure. If still too great, use a stronger light or try a longer exposure and use a very dilute developer. If still too great the negative is hopeless and should be reduced unless dodging will help it, as set out further on.

It will be noticed that this method calls for a one-minute development. This is desirable for several reasons: first, because it gives a unit and assists us in determining the correct exposure of other negatives, and second, because it is a comparatively short development, and yet gives sufficient time after the image has acquired the proper depth to pour off the developer and flush with water, thus stopping development. It also leaves sufficient margin in the event of over- or under-exposure. With one minute as the unit, over-exposure will result in a fully developed image in, say, thirty seconds. This print we could save; but if our unit were thirty seconds it would be extremely difficult to save a print which had completed development in fifteen seconds. The chances are that the development would go on to a ruinous extent before we could pour off the developer and flood the print, or that it would go on even after the water was poured on it. Moreover, in case of under-exposure, two minutes would not be so very tiresome, but four minutes would, besides which we would risk straining the print by such prolonged development. While I am not prepared to assert it as a rule, yet it has been my experience that the time of development varies almost inversely with the length of exposure; so that if the test-strip concludes development in half a minute with ten seconds exposure, I give the next five seconds exposure in the expectation that it will take a minute to develop. This assists greatly in lessening the number of test-strips required to ascertain the correct exposure of a given negative.

Should we wish to see a proof before the negative is dry, it is taken from the fixing bath and well rinsed, though not necessarily thoroughly washed. It is then placed face up in a tray of water, on which we place face down a sheet of bromide paper. The two are removed together and squeezed lightly into contact to remove air bubbles. The back of the negative is then wiped to remove superfluous water, and an exposure of several times the normal given, preferably the normal exposure at half the standard distance from the light. The paper is then removed and developed as usual. In this way it is possible to show a print in fifteen or twenty minutes after the exposure of the plate was made.

The purpose of the rinsing before development is to avoid the possibility of air-bells. The paper should be rinsed in cold water, as warmish water will cause air-bells instead of preventing them. This rinsing can be dispensed with if thought desirable. The rinsing after development is for the purpose of stopping development immediately, and also in order that the prints may not go into the fixing bath full of developer, as staining would be likely to result in such case. With the iron oxalate developer an acid rinsing bath is necessary, but it is not necessary with any of the other developers.

The fixing is important, as upon this depends in a large measure the permanence of the prints. The bath should be freshly made up, 3 ounces of hyposulphite of soda to 16 ounces of water. Prints are placed in this bath face down, and one under, instead of on top of another. The tray should be occasionally rocked. With a fresh bath prints will fix in ten minutes, but where many prints are made at one time it will be well to use a second fixing bath. The emulsion of an unfixed print will appear a yellowish tinge in the unfixed portions when examined by transmitted light; but this is not an easy or certain test. It is better to make absolutely certain of thorough fixing by continued immersion, occasional rocking and, where many prints are made, a second bath. The fixing bath should not be allowed to get too warm in hot weather. Blistering, staining and frilling will result in such a case, and I have known a print which was left in a warm fixing bath for an hour or more to be reduced beyond redemption. With freshly made hypo baths at a suitable temperature there is absolutely no danger of the paper frilling or blistering.

The final washing must be thorough, as the hypo is difficult to eliminate from both the emulsion and the paper. Care must be taken to see that the prints are well separated while washing. This ensures uniform washing.

It frequently happens that a negative may require more or less dodging in printing. With bromide paper this is particularly easy. We will take the simple case of a negative with dense sky which will not print out in the ordinary way. All that we need in this case is a piece of paper cut roughly to the sky line and kept moving during part of the exposure over the part which is to be held back. If necessary, cut down the light in order to prolong the exposure, or expose at a greater distance from the light. One or more test-strips will be required for this purpose in order to ascertain the relative times of exposure. A modification of this method is when a small portion of the negative only needs extra printing—a face or hand for instance. Here we take a piece of paper a little larger than the negative and cut a small hole in it, moving it in front of the light so as to throw the latter only upon the portions needing the extra printing. Still another modification is where a portion only needs holding back. Here we use a small piece of paper or cardboard stuck on a knitting needle, moving the latter so that it will not intercept the light too long at one place.

In all these and similar instances which will occur to the reader, the dodging should be done during the first part of the exposure. The subsequent exposure seems to obliterate traces of such dodging better than when it is done at the end of the exposure, just as in cloud-printing better results are achieved by printing the sky first and the foreground afterward.

It is quite possible to make bromide negatives in the camera. They have their advantages in classes of work not requiring the finest definition, are much lighter, cheaper, more easily stored and less liable to breakage or other mishaps. They are best made on a thin, smooth paper, a *soft* paper being better than the *hard*. They are placed in the plate-holder by means of the ordinary cut film holder. The exposure required is ascertained by a trial or two, but roughly speaking is about one-twentieth that of a rapid plate. After development in the usual way — it being carried only a little further than usual — and after fixing, washing and drying, the paper negative can be spotted or retouched, after which it is waxed.

Chapter IV
ENLARGING—DAYLIGHT METHODS

Contents

In taking up enlarging a full knowledge of what has been said as to the manipulation of bromide paper will be necessary, as the principles governing exposure apply here as in contact printing, errors being even more serious, owing to the greater waste of material.

For the illuminant used in enlarging, we may employ either daylight or artificial light. The former is cheap, but variable; the first cost of the latter is quite a little sum, but the light is uniform. A daylight enlarging apparatus can be made for a dollar or two, and hence is within the reach of all; and if the process be given up, the loss is not serious.

If the cost is of little or no moment, very serviceable enlarging cameras can be bought for about twenty-five dollars. Such a camera is adapted for reducing as well as enlarging, and so will be found useful for lantern slide making, copying, etc. As a matter of fact, few things are as useful to the amateur as a good enlarging outfit.

We will first consider enlarging by daylight with home-made apparatus. For this purpose a room with at least one window will be needed. It should preferably be convenient to the dark-room. If the window of this room commands a view unobstructed by buildings, trees or the like, so much the better. I personally prefer a south light. With this one can get soft enlargements from the most contrasty negatives, while by shielding the negatives from the direct rays of the sun we can work from negatives which are quite flat and lacking in contrasts.

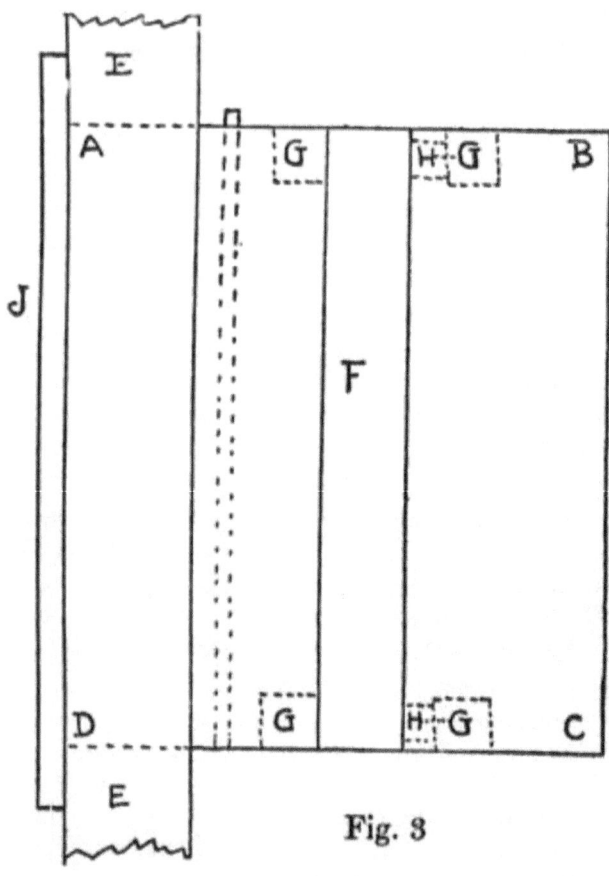

Fig. 3

But whatever the room chosen, all windows but the one at which we are to work must be blocked up. This can be done by heavy dark curtains, or by specially constructed frames covered with light-tight material and made to fit closely in the windows. If there are any transoms these should likewise be covered. White light entering under the doors can be shut out by placing a rug along the bottom of the door. Care must be taken that the window-frames fit closely, as the light from openings at the windows would soon fog a sheet of bromide paper if it fell upon it even for a few moments.

Assuming that the room chosen can be made practically light-tight, we will need some arrangement to hold the negative. The details of a box for this purpose can best be shown by a diagram (Fig. 3). ABCD is a strong and neatly made box open at both ends, and about two inches larger each way than the largest negative from which enlargements are to be made. E represents a section of a board which forms part of a window frame, a general view of which is given in Fig. 4.

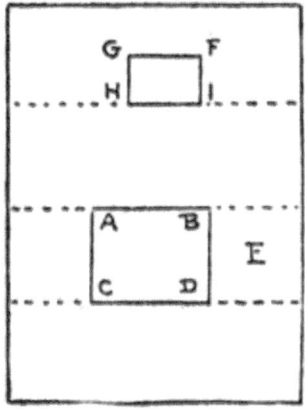

Fig. 4

Reverting to Fig. 3, F is an opening cut in the side of the negative box two inches or a little less from the back of the box, AD, and wide enough to admit the free passage of a negative in a kit or other holder. On the inside of the box are tacked strips, GGGG, to serve as a guide to the kit when placing it in the box. An opening similar to F should be made in the other side of the box to permit lateral adjustments when we come to use the apparatus, besides enabling us to put the negative in or withdraw it from either side. A convenient modification of the strips, G, is found by placing the front ones a short distance further forward, to wit, toward BC, as they are shown in the cut (Fig. 3), and tacking to them a piece of watch spring, H, this then serving both as a guide and as a means of pressing the kit or negative holder against the other strips, GG (Fig.

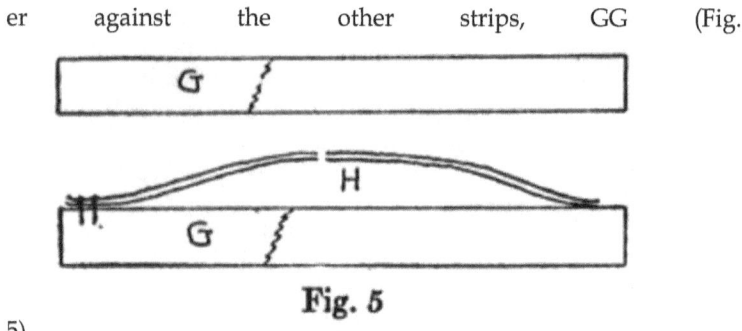

Fig. 5

5).

J is a sheet of ground-glass, which is tacked over the opening when the box is firmly set in the board, E. It is well to have this ground-glass fixed in place so that it can be readily removed if desired.

The necessity for having the box at least two inches larger each way than the largest negative from which enlargements are to be made is shown in Fig. 6.

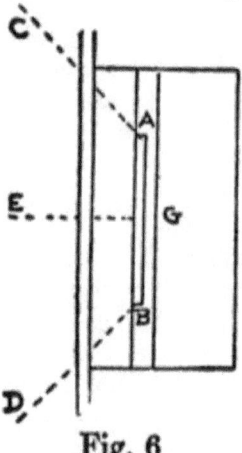

Fig. 6

Here AB represents the negative in place, CA, DB and EG represent rays of light entering the box. It will be seen that the rays CA and DB strike the ground-glass at an angle, but nevertheless at an angle which results in their passing through it in a considerable degree. They strike the negative AB, but if the negative were the full size of the box, to wit FG, it will be seen that while the section AB would be fully lighted, the sections AF and BG would receive no oblique rays at all, and hence the negative would not be even approximately uniformly lighted. This point is too often overlooked in the construction of apparatus of this character, but is necessary in all cases of daylight enlarging and especially when direct sunlight is used. Now with the negative box in place, some arrangement must be made for holding the lens, which can be the lens used for making the negative. This for enlargements of a fixed size from negatives of a given size can be accomplished by simply extending the section BGGC Fig. 3, to a proper distance and placing the lens in the end of it; but this permits too little opportunity for adjustments and is not advisable. A double box, one sliding within the other, would be better, but still not quite satisfactory. It is far better to adapt one's camera to the apparatus, and this can always be done; it being very simple with a reversible back camera, which can be backed right up to the opening, and more difficult but always possible with others. Fig. 7 shows the entire apparatus in place, and the manner in which it is used. AB is the window board, C is the negative box, D is the camera adjusted to the latter, E is the enlarging screen on an easel to hold the bromide paper, and F is the reflector. The screen on the easel can be made either to rest on the floor or on a table. It can be made to run on a track or otherwise, and it can also be made so as to admit of either vertical or lateral adjustment or both, or it can be nothing more than an ordinary box set on a table. But however constructed it must be considerably larger than the largest sheet of bromide paper which is to be used, thus allowing for nearly all necessary adjustments of the paper. It is preferably covered with white paper or fine blotter to aid in focusing. The reflector F is considerably larger than the negative-box, and adjusted at an angle which will reflect the light from the sky or sun evenly upon the ground glass. It is best covered with good white blotting paper. G is a hood which I have found useful in sunlight enlarging, especially in summer when the sun is almost overhead. It is placed on the outside of the window-frame, some distance above the ground-glass, and shields the latter from the direct rays of the sun, which would otherwise cause uneven illumination owing to their too great obliquity. The direct sun on the white reflector will give a light of high intensity. In winter, however, when the sun is low, it will fall directly on the ground-glass, and this, if the negative box be constructed as advised, is not objectionable, but on the contrary an advantage. In Fig. 4 the opening, FGHI, represents a sheet of ruby glass, and can be screened while focusing if found to interfere with the worker's convenience in that operation.

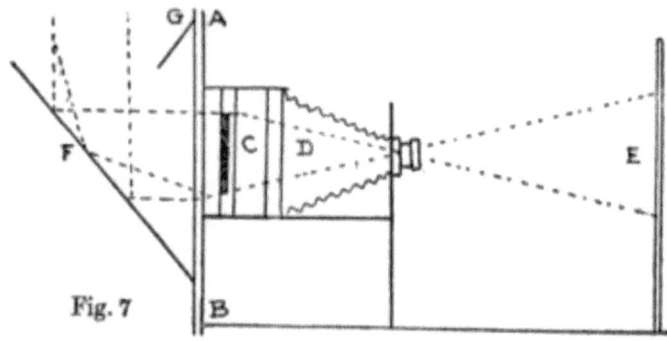

Fig. 7

The apparatus as sketched will suffice for all ordinary work. Modifications of it will depend upon the ingenuity of the man who attempts to design or construct one. It should be noted that the distance of the ground-glass from the negative has its influence in the strength of the light, and it is better to have this distance not over two inches. If less than one inch, however, the diffusion of light is not so good. When the light is weak the ground-glass can be removed entirely; the negative will thus be viewed directly against the white reflector. Very strong negatives giving undue contrasts may also be dealt with in this way. Or, if the light is too strong for flat negatives, the reflector can be removed entirely, or to the same end a sheet of yellow glass can be substituted for the ground-glass, thus increasing contrasts. In fact, a very useful and easily arranged modification of the negative-box consists of an opening in the top of the box inside the room through which can be dropped an extra sheet of ground-glass or opal to cut down the light, or of yellow glass to increase contrast. This opening should be at the point K, Fig. 3.

I have referred to a kit as being the proper arrangement for holding the negative. This, after much tribulation in working with home-made contrivances, I have found to be the best arrangement. They come a size or two larger than the negative with which they are to be used, and can easily be cut down to the proper dimensions. With it, also, other kits to hold smaller negatives can readily be used. It is also simple with them to fasten the negatives in place. If they extend beyond the box on either side so much the better; greater lateral adjustment can then be made. The negative box, Fig. 3, is best painted dead black inside in the section GBCG, and white in the section AGGD. The reasons for this will be obvious at a glance.

In enlarging from films it is well to place them between two sheets of glass of proper size, and fasten the whole in the kit or negative-holder. For this purpose use thin glass without flaws or scratches. If the films are smaller than the opening in the kit, it is well to paste a black mat on one of the glasses, when, after proper adjustment, the film will remain in place between the two glasses with very little pressure.

Enlarged negatives are very easily made with the apparatus described. A contact positive can be made, preferably on carbon transparency tissue, and from this the enlargement made, or an enlarged positive made first, and from this a contact negative. The latter plan is preferable, since it admits of retouching on both positive and negative. Slow plates should be used throughout. For those who do not care to go to the expense of experimenting with large plates, I would suggest that good contact positives be first made and from these negatives on bromide paper, *Standard A, soft*. These negatives are treated as already described. The best positive for this purpose is a thin one with full gradations of tone from clear high light to deep shadow, without veil or fog, but free from any suspicion of flatness.

Chapter V
ENLARGING BY ARTIFICIAL LIGHT

Contents

The apparatus for enlarging with artificial light is, as has been stated, more expensive than that for use with daylight. The negative box and screen, however, remain as given. But we need in addition two extra pieces, a light-box and a pair of condensing lenses.

The form of light-box presupposes the choice of illuminant, and in this there is a wide range. Suffice it to say that a kerosene lamp with one or more one-and-a-half inch burners will be found suitable for very small work or weak negatives. For larger work or stronger negatives a stronger light will be needed.

Of these, the first in point of strength is the arc-light, which is too strong for ordinary negatives to be enlarged not more than fourfold on ordinary bromide paper. Used with any of the slower papers it will be found very serviceable and satisfactory. Next comes the lime-light, which has pretty much the same advantages and disadvantages. After these come acetylene, a gas giving an intense light of high actinic power. This is within the reach of nearly all, as a first-class generator costs only about twelve dollars, and the uses of the gas are manifold. The same generators and burners can be used with a projecting lantern and will be found far more satisfactory than oil. Acetylene burners can be had in various sizes, ranging in power from thirty to several hundred candle-power. The carbide from which the gas is generated is not expensive and costs only a few cents each time the machine is loaded. By an adjustment attached to the generator the gas is kept at a constant pressure, and hence the light is unusually steady. All in all this light has many advantages. After it in strength comes the Welsbach burner, suitable for those having gas in the house. After this comes the ordinary gas-burner, and then oil. The reader, knowing now what will be required of his light, can take his choice.

Perhaps the simplest form of light-box is where the light is placed in one room and the enlarging done in an adjoining one, the light being admitted through a suitable opening. This prevents the possibility of stray light reaching the paper and is productive of no additional heat in a room presumably already hot enough.

If this is not feasible a light-box must be constructed. As these vary so much in material and design, and must be altered with different forms of light in use, I will merely state the requirements. First of these is that it must be light-tight, and second, that it must have adequate ventilation and be fire-proof. Following these in importance, there should be a simple arrangement for looking at the light from time to time to see that it is burning properly and some means for readily attending to it if it is not.

Having the light-box, the burners must be placed in it properly. Here the shortest way out of the difficulty is to go to an expert. If electricity is used go to an electrical supply house; if gas, go to a gas-fitter. As will be seen later the flame itself must be placed in a certain relation to other portions of the apparatus, and provision must be made accordingly.

In looking over the magazines and annuals we will now and then see some new method given for illuminating evenly the back of a negative in enlarging or reduction. The most of these the writer has tried, but he has never found one of them which could be relied upon to give even reasonable satisfaction. If the light is apparently evenly diffused it is too weak. If strong enough it is not evenly diffused. Hence I will recommend nothing short of a pair of condensing lenses, as these have been proved by experience to be satisfactory in every respect if properly handled and cared for. The diameter of these must be slightly greater than the diagonal of the largest negative from which enlargements are to be made. These can be bought in pairs, mounted or unmounted, at about the following prices:

Diameter Inches	Focus Inches	Pair of Lenses Mounted	Per Single Lens Unmounted
4	5½ or 6½	$ 4.00	$ 1.25
4½	5½ or 6½	6.00	1.50
5	6½	7.50	1.75
6	8	12.00	3.00
8	12	32.00	7.50

Diameter Inches	Focus Inches	Pair of Lenses Mounted	Per Single Lens Unmounted
9	14	40.00	10.00

The prices asked for condensers vary considerably in different price-lists. They can often be had at second-hand at a decided saving of expense.

If it is desired to save the additional cost of the mounted condensing lenses, they can be comparatively easily mounted by anyone at all familiar with tools in the following manner:

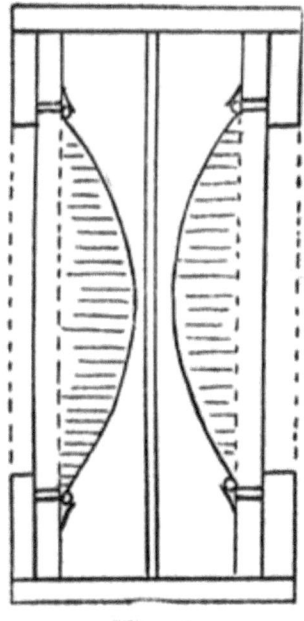

Fig. 8

A piece of quarter-inch pine or poplar is cut to a square about an inch larger than the diameter of the lenses. In the center of this is sawed out a circular opening the exact size of the lens. In another board of the same dimensions is cut a circle a quarter of an inch less in diameter. These boards are placed together with the grain running in opposite directions, to prevent warping, and the lens kept in place by a wire bent in a circle and clamped in place so as to hold the lens, or other similar arrangement. See Fig. 8. The other lens is mounted in the same way. The two are mounted with their convex sides facing each other and a slight distance apart. It is better to place between them a thin sheet of finely ground glass, as this overcomes the bad effects of slight flaws in the lenses, which are not uncommon. The combination is then boxed up.

Having now our light-box, condensers, negative box, camera and screen, they are next arranged in the order shown by Fig. 9. A long table especially constructed for the purpose makes the best base for this purpose.

In putting the apparatus together there are several points to be noticed. First, the planes of the lenses, negative, projecting lenses and screen must all be parallel; second, the centers of all these should be in a single straight line, and third, either the light or the condensers should be so mounted as to easily slide backward or forward, since every time the projecting lens is racked backward or forward it necessitates a corresponding motion of the condensers to or from the light.

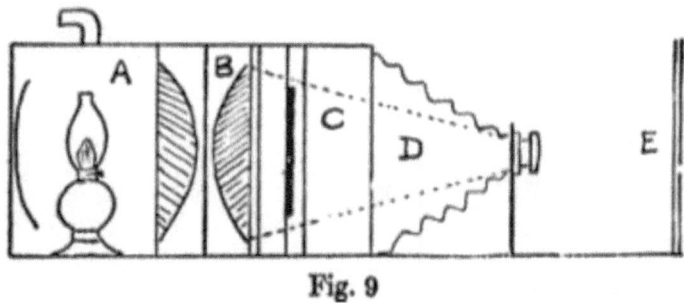

Fig. 9

In constructing the apparatus, for use with condensers and artificial light, the same provision should be made in the negative box for inserting a piece of colored or ground glass as was made in the daylight apparatus. When the diameter of the condensers is but little greater than the diagonal of the negative it will be necessary to have the latter quite close to the former, as the cone of light from the condensers has its apex at the lens, and hence if the negative in such a case is at a distance from the condensers the corners will receive no light. Reference to Fig. 9 will show this plainly. For this and other reasons it is always best to have the condensers of ample size for a given negative. In fact, before beginning to make enlargements the worker should work with one good negative until he finds out exactly what light-intensity is best suited to it. This will then serve as a standard for all other negatives of the same general grade, and variations of the light can be made as required for particular negatives, or where the extent of enlargement is materially changed for various purposes.

In using the daylight apparatus, which we will now consider, the negative is placed in the holder opposite the center of the ground-glass, upside down and facing into the work room. The room is darkened and lens uncapped. An image more or less blurred will appear on the screen. If the enlarged picture is to be only slightly larger than the negative, the lens must be racked out until its distance from the negative is but little less than its distance from the screen. To make the enlargement greater we simply rack back the lens and move the screen further off. There are tables which show exactly the distance which the lens must be from the negative and screen in order to get an enlargement of a given size: The table here inserted covers the ordinary requirements and may be of service in constructing the apparatus:

TIMES OF ENLARGEMENT

Total distances from the negative, in inches.

Times of Enlargement	2		3		4		6	
Focus of Lens	To Easel	To Lens	To Easel	To Lens	To Easel	To Lens	To Easel	To Lens
6 inches	27	9	32	8	37½	7½	49	7
8 "	36	12	42⅔	10⅔	50	10	65⅓	9⅓
10 "	45	15	53⅓	13⅓	62½	12½	81⅔	11⅔
12 "	54	18	64	16	75	15	98	14

A table for enlargements of from one to twenty-five times, with lenses varying in focal length from three to nine inches is here given.

ENLARGEMENTS

From the British Journal of Photography Almanac.

Focus of Lens	Times of Enlargement and Reduction							
Inches	1 inch	2 inches	3 inches	4 inches	5 inches	6 inches	7 inches	8 inches
3	6	9	12	15	18	21	24	27
	6	4	4	3	3 3/5	3	3/7	3 3/8
3	7	10	14	17	21	24	28	31
	7	5	4 2/3	4	4 1/5	4 1/12	4	3 9/10
4	8	12	16	20	24	28	32	36
	8	6	5 1/3	5	4 4/5	4 2/3	4 4/7	4
4	9	13	18	22	27	31	36	40
	9	6	6	5 3/5	5 2/5	5	5 1/7	5 1/16
5	10	15	20	25	30	35	40	45
	10	7	6 2/3	6	6	5 5/6	5 5/7	5 5/8
5	11	16	22	27	33	38	44	49
	11	8	7 1/3	6 4/5	6	6 5/12	6/7	6/10
6	12	18	24	30	36	42	48	54
	12	9	8	7	7 1/5	7	6 6/7	6
7	14	21	28	35	42	49	56	63
	14	10	9 1/3	8	8 2/5	8 1/6	8	7 7/8
8	16	24	32	40	48	56	64	72
	16	12	10 2/3	10	9 3/5	9 1/3	9 1/7	9
9	18	27	36	45	54	63	72	81
	18	13	12	11 2/3	10 4/5	10	10/7	10 1/8

The object of this table is to enable any manipulator who is about to enlarge (or reduce) a copy any given number of times to do so without troublesome calculation. It is assumed that the photographer knows exactly what the focus of his lens is, and that he is able to measure accurately from its optical center. The use of the table will be seen from the following illustration: A photographer has a *carte* to enlarge to four times its size, and the lens he intends employing is one of 6 inches equivalent focus. He must therefore look for 4 on the upper horizontal line and for 6 in the first vertical column, and carry his eye to where these two join, which will be at 30-7½. The greater of these is the distance the sensitive plate must be from the center of the lens; and the lesser, the distance of the picture to be copied.

In practice it is convenient, after having once found the focus for a given enlargement from a given negative with the lens in use, to mark on the base of the apparatus the point to which the lens has been extended. Then in making future enlargements of the same size, it is only necessary to set the lens at that point and move the easel backward or forward until an approximate focus is obtained, when the image will be of the proper size on the screen.

As an approximate guide it is sufficient to know that the nearer the lens is to the negative the greater will be the enlargement, as may be seen in Fig. 7. If a piece of thin cardboard, or a sheet of paper cut to the exact size of the enlargement desired, is placed upon the easel-screen, little difficulty will be experienced in getting an enlarged image of the proper size and correctly focused.

It is advisable to focus the enlargement with the largest aperture of the lens. If the lens, when working at its largest aperture, covers the plate from which the enlargement is being made, it will give proper definition over the enlargement. With a lens of the better sort, of course, the definition will be equally good whether a large or small aperture is used; but with a low-priced lens it is better to stop down to No. 8 ($f/11.3$) or No. 16 ($f/16$), to avoid spherical aberration. Stopping the lens down increases the time of exposure, and enables one to have greater control over the operation of exposing the paper, permitting time to shade or locally increase the exposure at any portion of the image. This is sometimes useful, but as a general thing stopping the lens down is not advisable, as interfering with one's judgment in calculating exposures for various negatives. Having secured the image correct in size and focus, place thumbtacks at all four sides of the sheet of paper or card used to focus the image. These will serve as a guide to the placing of the sensitive paper. Adjust the lens stop as desired and cap the lens, leaving the room totally dark save for such safe light as we may have for working. Place the bromide paper on the screen, using the thumb-tacks as a guide to the correct position in this.

In making his first enlargements, the beginner should avail himself of the help of test-strips. These should be about one inch wide and the length of the paper. The exposure depends on a number of factors, among which are light, negative, focal length of lens, size of enlargement, stops, sensitiveness of paper, developer, temperature of developer, and length of development. The first experiment had best, therefore, be made on a purely arbitrary basis, for which we will take ten seconds.

Pinning a test-strip on the screen, we uncap the lens and with a piece of cardboard shade two-thirds of the strip during five seconds; move the cardboard, and give the next section five seconds making ten for the first; then remove the cardboard entirely and give five seconds more, making fifteen for the first, ten for the second, and five for the third. Now develop the strip. If the fifteen seconds portion finishes development in less than one minute, and the ten takes approximately a minute, we will know that our basis was correct. But if all three were over-exposed or under-exposed, as shown by one minute's development, we can expose the next test-strip accordingly.

In determining the correct exposure, the method already set forth for contact exposures is a reasonably good one. If the paper with a given exposure takes half the proper time to develop, halve the next exposure; if double the time, or more, double it. More could be said on the subject of exposure, and possibly to advantage; for instance, there are tables showing the exact relation of exposure to the number of times of enlargement, but complicated calculations in the dark-room are troublesome and a test-strip is simpler. After a while one gets the ability to determine the approximate exposure required by looking at the enlarged image on the screen, correcting slight errors by length of development, and greater ones by modifying the developer by diluting or strengthening.

It should be remembered, however, that in judging exposure by reference to the screen, we must consider the high lights, as well as the shadows. It is in the high lights that we need the detail if we are to have soft pictures. If this detail in the high lights is plentiful and clear we may know that our light is strong enough for a very short exposure. If it is very faint, we will have to give a long exposure and use diluted developer to save the over-exposed shadows. On the other hand, if the image on the screen is a flat one, we may know that our light is too strong for the negative, and it must be modified by removing the reflector or by interposing ground or yellow glass; and if neither of these suffice, we can simply lay the negative aside for a dark day when the light will be very much weaker. Frequently all necessary contrasts can be obtained by the use of the *hard* paper before referred to. As under-exposure tends to increase contrasts, we should also give the minimum exposure in the case of flat negatives, abandoning for a time our standard one-minute development. As will be seen by this time, there are many wrinkles about using bromide paper, and it will be found that new ones appear at every *seance* in the enlarging room.

But why is it that so many of our enlargements are black in the shadows and chalky in the high lights? Why, simply because our light is too weak for our negative. We forget that if we cannot modify our negative we must modify our light. It is this characteristic of the bromide enlargement which has prevented the process from enjoying the popularity it deserves. And I sometimes wonder whether this chalkiness is due to the use of the north light!

Chapter VI
DODGING, VIGNETTING, COMPOSITE PRINTING AND THE USE OF BOLTING SILK

Contents

Of all printing processes, bromide enlarging offers the best opportunities for successful dodging and modification. We can cut our light down and take all the time we want, or we can take as little time as we want. A hand, a finger, a slip of paper, or anything within reach, will suffice to shade the light just as we want it. In this connection it is well to always hold the shade nearer the lens than the easel, as greater diffusion results and there is less danger of sharp lines. In shading a foreground to bring up a dense sky, first make a test-strip or two, noting how long the shading is carried on and how long the light is allowed to act on the whole. If the sky is then over- or under-printed we can modify it in the enlargement proper.

The best arrangement for vignetting in enlarging is a piece of cardboard the size of the negative, with an opening cut out at the proper place and about the size of the portion of the negative to be vignetted. This is held near the lens and moved backward and forward between the latter and the screen, the opening showing larger as we near the lens and smaller as we recede from it. Very tasteful vignettes can be made in this way. A favorite method of the writer's is to use a sheet of bromide paper, preferably that with rough surface, and print on it a small vignette of a portion of a negative. These sheets being of a uniform size are then bound in book form, and make very attractive souvenirs. Variety can be added to the collection by printing some of the pictures through a mat fastened on the screen over the paper, when, of course, they are bounded by straight, sharp lines.

Double printing in enlarging is not at all difficult. Assuming that test-strips have been made determining the proper exposure for each negative, I will briefly outline the process. Taking a landscape negative with clear sky in which we wish to print clouds, we first tack on the screen a sheet of paper the size of our bromide, and after properly adjusting and focusing it, trace with a pencil the outline of the skyline. We then remove the foreground negative and, after tracing, cut out a mask conforming to the size and shape of the foreground, cutting away the sky. We now put in the box the sky negative, and readjust our sheet of paper until after proper focusing the desired portion of the sky occupies the portion reserved for it, leaving the thumb-tacks as a guide when we put our bromide on the screen. Now using the sheet of paper as a guide, place on the edges of the bromide paper two little pencil marks to show how far we shall shade the lower portion of the paper. Our mask being the size of the foreground negative, it is now only necessary to hold it at the proper distance from the lens to have its edge conform to the sky-line when enlarged. But this would leave a sharp line if held exactly at that point, so using the pencil marks on the margin as a guide, we slowly raise and lower the mask very slightly and just sufficient to cause an agreeable blending of the sky into nothing. The proper exposure given, we cap the lens, remove the paper and insert the foreground negative. Now we must again adjust our sheet of plain paper until the sky line marked on it coincides with the sky-line on the screen, leaving thumb-tacks as usual. Registry being thus secured, we simply expose the foreground and develop the composite print.

Needless to say, our clouds must be lighted from the same general direction as the landscape. But if in the negative they are not so lighted it can be reversed in the holder and will then print properly. In almost all cloud printing the tendency is to give undue prominence to the clouds by printing the sky to too deep a tone. This, besides making the blending very noticeable at the horizon, results in unnatural effects and should be avoided.

If the sky portion of the landscape negative is thin, it might print slightly and spoil the effect of the clouds. This can be overcome by using a weaker light in enlarging. Where this is not desirable, a mask can be cut for the sky portion and used slightly while the foreground is being printed. By using it a very little during the first part of the exposure the tint will be overcome, while objects projecting above the horizon will be sufficiently printed. It will be found easier, no doubt, to print the landscape first and sky afterwards. But this does not result in good work. The landscape should invariably be printed after the sky portion.

Bolting silk enlargements were for a time very popular. Sometimes they were artistic. Then every-one began making them, too often from unsuitable negatives, and they fell into disrepute. This method of enlarging is, in fact, suitable for very few negatives and only where broad effects of light and shade are desired. To cut up a spotty negative with a succession of lines does not necessarily give a broad effect in the picture. But for softening down heavy masses of shadows, and blending them harmoniously with masses of light or half light, the process is without an equal. The bolting silk can be bought by the square yard of dealers in photographic supplies, and should be stretched evenly over a frame made of quarter- and half-inch wood, being tacked between the two strips. This frame can be easily adjusted to fit over the paper on the screen. By using the side, bringing the cloth within a quarter of an inch of the paper, the lines are more evident, which is not so objectionable for very large work. But for the softest effects, reverse the frame and use it at half an inch from the paper. In this way we get a soft diffusion of the lines and much greater general softness. It should never be used nearer than a quarter of an inch, as the lines then become too evident, and hence fail in the effect desired. The bolting silk comes in three grades, fine, medium and coarse. The medium is the best for general work. It should not, however, be used for sizes under 8 × 10. The interposition of the cloth requires about one-half additional exposure. Focusing, of course, must be done without the frame in place. The bolting silk should only be used with paper which is to be toned to some color other than black, as there is something incongruous in its use with black tones.

Few branches of photographic work, outside of negative-making, are as fascinatingly interesting as the making of enlarged prints on bromide paper from small negatives. Every amateur has negatives worthy of enlargement in his collection, and the process is so simple as to be within the capacity of the amateur who is still in his first year in photography. Its practice will stimulate his interest and help him in all his other photographic work. Especially will it help him in picture-making, the merits and defects of composition being a hundred fold more plainly evident in an enlargement than in the small print from the hand-camera negative. Moreover, in its essentials, bromide enlarging calls for no special equipment other than the ordinary hand- or view-camera, and a dark-room or other convenient work-room from which all "white light" can be excluded on occasion.

Chapter VII
THE REDUCTION AND TONING OF BROMIDE PRINTS

Contents

The subsequent manipulations with bromide paper do not differ materially from those with negatives. The support being paper of course makes some difference and the fact that while in the negative we aim to get printing density and printing color only, in the positive we aim to please the eye, makes another. But generally speaking, it may be said that whatever we can do with the negative we can do with bromide paper, that is, in so far as the emulsion itself is concerned.

The first operation to be taken up is the reduction of prints which are too dark. This can best be effected just after the prints come from the hypo. A few grains of red prussiate of potash are dissolved in a suitable quantity of water, the latter being barely tinged, not of a strong yellow color. If the print is too dense throughout, it can be immersed without previous washing in this solution. Reduction should take place gradually, and this is best accomplished with a weak reducer. If the tray be rocked gently the reduction will be quite uniform. If, however, only a portion of the print needs reduction, this can be effected by applying the ferricyanide solution locally with a brush or bit of absorbent cotton. Extreme care is needed in this operation. In this way unduly deep shadows can be softened, veiled high lights brightened, or almost any modification obtained which may be deemed desirable. When reduction is almost completed quickly rinse the print in running water and then wash thoroughly. If the print has been dried, it is only necessary to soak it for a few moments in a fresh fixing bath, when the ferricyanide can be applied as before.

Latterly the toning of bromide prints has become popular. There are many methods and innumerable formulae. Here we shall concern ourselves with the sulphide method which best fulfills the three chief requirements, namely: (1) Certainty of results; (2) the use of few baths; (3) the production of permanent prints. Processes which, as regards color, vigor, etc., are beyond the control of the worker, are of very little practical use. Equally so, if the toning involves a whole string of operations, the final outcome of which is usually—a spoilt print. And, lastly, a process which—however satisfactory it may be in other respects—impairs the undoubted permanency of a black-developed print is not one worthy of adoption. In one or two other respects, processes vary chiefly as regards the depth or intensity which the print must have in order to produce the most satisfactory result when toned. Thus, prints to be toned with uranium require to be distinctly on the pale side, whilst those for sulphide toning are best a little vigorous. One or two other methods, on the other hand, require the use of the costly gold or platinum salts. The latter, except under exceptional circumstances, are far better employed in the legitimate form of platinotype or other platinum paper; bromide prints toned with platinum will probably cost more, and will never have the absolute permanence peculiar to the platinum print.

Placed in rough order of merit, the processes available are: Sulphide toning (hypo-alum toning is a cheaper, slower, and not quite so effective form of this method, whilst the thio processes represent sulphide toning at its best); copper toning; toning by re-development. These methods differ, not only in the results which they give, but also as regards the perfection with which each attains its particular effect; on the principle of the lady in the play who spoke the "absolute truth under the circumstances," each may claim to be included among the really serviceable processes.

In the sulphide process, the image which, in a black-developed print, consists of metallic silver in fine division, is converted into silver sulphide, a substance which in the ordinary way is also black, but when produced in a fine condition on a photographic print is brown to sepia color. Silver sulphide is a most permanent substance. Therefore a sulphide-toned print should be permanent, too, a conclusion which is fully borne out in practice. A sulphide-toned print is at least as permanent as the bromide from which it is made. The image of the latter is susceptible to practically only one agent likely to come in contact with it, namely, sulphur fumes from burning gas, which partially sulphurize it and give rise to iridescent markings resembling those due to stale paper. Now, as the sulphide-toned print is the result of this sulphurizing process carried out with intention to a state of completeness, the result should be—and proves to be in practice—immune to this one cause of defacement.

In converting the silver image into one of silver sulphide, the method is to first act on (bleach) the silver image with some reagent which will change it into a compound of silver susceptible to the action of sulphide. Iodine has been used for this, giving an image of silver iodide. Bromine gives one of silver bromide. A mixture of potass bichromate and hydrochloric acid gives silver chloride, as does also a solution of chlorine, though in the former case the presence of the chromium compounds affects the color obtained. But the best of the lot is a solution of the two substances potassium ferricyanide and potassium bromide, which forms an image of silver ferrocyanide and silver bromide. Both of these are converted into silver sulphide when treated with a solution of sodium sulphide. In the case of the hypo-alum process, in which the prints are toned direct (without bleaching) in a mixture of hypo and alum, the image is also changed into silver sulphide, but only to a partial extent. Theoretically, the method is not so good as sulphide proper; it is much more inconvenient in practice except on a commercial scale, while the results cannot be said to quite equal those by the sulphide process as regards permanency.

So much by way of theory. We will now give the formulae for the two solutions required in the sulphide process. The first of these is the "bleach," or oxidizing mixture of bromide and ferricyanide. Within reasonable limits, the proportions of these salts and the quantity of each in the solution does not matter very much. Each chemical can, if desired, be kept in a separate solution if care be taken to keep the mixture in the dark,—that is, in a cupboard where it will not be exposed constantly to daylight. The ferricyanide suffers in time by exposure to daylight; but, as both it and the bromide are comparatively cheap and serve for a large number of prints, there is no need to take excessive care. The ferricyanide-bromide mixture, however, keeps very much better than a plain solution of ferricyanide alone. Formulae which place the salts in separate solutions are a mistake.

As good a formula as any is: Potass ferricyanide, 300 grains; potass bromide, 100 grains, water 20 ounces; Ammonium bromide may be used in place of the potassium salt in the above formula; the difference is not marked, but the ammonium compound tends to give a somewhat warmer brown or sepia. In the case of many formulae, it will be noticed that equal quantities of bromide and ferricyanide are recommended. Although, as just stated, variations in the formula are not at all marked in their effects, a proportion of bromide over one-quarter of the ferricyanide does tend towards the yellowish color of which complaints are now and again heard. I want to make it clear that the opportunities for going wrong with the bleacher are very small indeed. Without encouraging the reader to be careless let it be said that "any old formula" (of ferricyanide and bromide) for the bleacher will prove successful. Not so, however, in the case of the sulphide solution, which requires to be very carefully made up and used.

Sulph*ide*, not sulph*ite*. The material for the toning or darkening of the bleached print is the chemical substance, sodium sulphide, of the formula Na_2S9H_2O. This is purchased as small crystals which greedily absorb water and rapidly become almost liquid if not properly corked. Not that this totally unfits the sulphide for use. Sulphide which has gone liquid will at all times be found to work perfectly, but it is of course open to suspicion, and, in any case, it is not possible to know what is the strength of a solution made up with such a supply. For this reason, it is best to make up the sulphide into solution of 20 per cent strength, and add this to water to make the toning bath. And it is here that a caution must be noted. The weak working solution, which is only about 1 to 2 per cent strength, keeps very badly indeed, and should be made up fresh from the stock solution at the time of toning each batch of prints. This is one of the most necessary items to bear in mind in using the sulphide process.

Sodium sulphide is sold in various degrees of purity, and the label on the bottle is not always in exact correspondence with the condition of the substance inside, but the two forms which must be adhered to for sulphide toning are the ordinary "pure" and the "pure for analysis." The former can be obtained from any reliable drug store or photographic dealer. It comes in small lumps, yellowish to greenish in color; when dissolved in water the solution will be yellow, and will usually show a deposit which must be filtered off. This sulphide will give tones which are sepia brown with most papers. In the case of the "pure-for-analysis" sulphide, which is the recrystallized variety, the salt will be pure white and will form a quite colorless and clear solution in water. The tone given by this kind of sulphide is usually of a more purplish color. The distinct difference between the two commercial varieties of sulphide should not be overlooked, as it allows the worker to modify the process usefully when dealing with papers differing (as all papers do) to a slight extent in their adaptability to sulphide toning. The purer form has certainly much better keeping properties than the other, but either, if made up in 20 per cent solution, keeps for a month or two at least—which is enough for all purposes. The chief difference between the two is noticed in the diluted or working solutions. That of the purest sulphide *may* be kept and used again, though it is not really good policy to do so.

The supply of sulphide should therefore be dissolved as soon as purchased, as follows: *Stock sulphide solution*—20 per cent; sodium sulphite 4 ounces; water to make 20 ounces. The actual toning solution is made up at the time of treating the prints by mixing the above stock with water, as follows: *Sulphide toning bath.*—Stock 20 per cent solution 3 ounces; water to make 20 ounces.

This makes a bath which contains about one per cent real sulphide, corresponding with about a 3 per cent solution of the sulphide as purchased. If the bath is much weaker, the tone obtained is usually not quite so good; while, if it is stronger, there is danger of the print's blistering while toning, or afterward in the washing water. Indeed, some papers need to be toned in a weaker bath, and require also to be fixed in an alum-hypo fixing bath (see later), so that the strength of the toning bath given above may be taken as the maximum, and used at half or one-third strength, as circumstances show to be necessary. And, to repeat the caution once more, the toning bath is to be thrown away as soon as the prints have been passed through it. With these points in our mind as to the making up of the solution, we can come to the process proper.

The prints require to be well washed and free from hypo before being placed in the bromide-ferricyanide bleacher, because any hypo in conjunction with the ferricyanide will form the well-known Farmer's reducer, and cause patchiness of the prints. It is immaterial whether the prints are taken direct through the toning process or dried in the meantime. Some workers contend that the toning process is more regular if the prints are dried before bleaching. In either case, immersion in the bleacher will cause the fully developed bromide to disappear, leaving only a faint brown image behind. In some cases the image is fainter than in others, the difference appearing to depend chiefly on the developer employed. Developers with a liability to stain will give prints which do not bleach out so completely as those made with cleaner working developers. But, in all cases, two to three minutes' action of the bleaching solution will be ample; if all pure black is not gone in this time, it is a sign that the bleach is becoming exhausted. The prints should be kept constantly on the move whilst in the solution, and turned over and over to ensure equal action. They are then given quite a brief rinse in running water — half a minute to a minute — and then transferred to the sulphide solution, where they should darken to the full brown or sepia tone in a few seconds. It is well always to leave them here for twice to three times the period required to give the full tone. A wash of half an hour will remove the salts left in the film.

Granted that bleacher and sulphide are in proper working order, there is one further factor in the making of sepia prints which is of vital importance, and that is the proper preparation of the print itself. A good sulphide tone presupposes a good black and white bromide. Not only that, defects in the bromide which may lie latent while the print is untoned come to light in the sulphide bath. This applies to uneven fixation (due to omission to keep prints moving in the hypo bath) and fingering of the surface; while, as regards the original development of the print, making the best of a wrong exposure will not do when sulphide toning is in view. A print that is forced by long development will suffer in tone, the result being colder and less satisfactory as regards vigor. Full exposure, and development which is complete in the normal time for a perfect black print, are the conditions for a good sepia tone, and, when a batch of prints is being put through, it is well to take steps to preserve a uniform time of development in order to secure an identical tone throughout.

There are many different formulas for the uranium toning of bromide prints, and I suppose that most of them have given good results with the workers who published their methods. Of those which I have tried, however, none has yielded the results which I have been enabled to obtain from my own formula—my own in that I arrived at it by patient experimenting. It may be that this formula is not wholly original with myself. At any rate, I do not claim anything for it except that it works, with me, better than others I have tried.

The requirements for toning bromide prints with uranium are: 1 ounce of uranium nitrate; 1 ounce of potassium ferricyanide (the red crystals); ½ pound bottle of acetic acid—c. p. glacial preferred; water; a supply of blotting paper, to be kept exclusively for this purpose, and a few absolutely and chemically clean trays.

The expense attached to these toning processes is slight. Uranium nitrate costs from forty to sixty cents per ounce, and an ounce will last a long time. Potassium ferricyanide costs about twenty cents per pound, and a pound is ample for a lifetime. Glacial acetic acid is a little more costly, but a half-pound bottle will prove a good investment. It is used also, as the reader will recall, in making acid hypo for acid fixing.

To prepare the toning baths, dissolve the ounce of uranium nitrate in 10 ounces of water. The water should be distilled if this is easily obtainable, and the solution should be kept in an orange-glass bottle or an ordinary bottle protected from light by a non-actinic paper wrapping. Dissolve the ounce of potassium ferricyanide in 10 ounces of water. Keep this also in an orange-glass bottle, well corked. There are many cautions about this particular salt, and it has been said that it will not keep in solution. In my practice I find no difficulty whatever in the use of a solution six months old, despite the difficulties mentioned in the text-books.

To tone the bromide prints, first note that the prints should have been developed and fixed and washed just as usual. It is necessary that prints to be toned shall contain no trace of hypo. To secure this, the prints should be specially prepared for toning by being again thoroughly washed, as any hypo remaining in the print will cause spots and streakiness. With care at this stage the toning will give clean and bright prints, which should be as permanent as the original bromide print.

I cannot give the reason why, but, as a general rule, bromide prints tone better if the print has been dried after washing and re-wet just before toning. There may be a chemical reason for this, but I am inclined to think that it is a physical one, viz., that the emulsion is softer after its first washing than after having been dried and wet, so that it allows toning solution to get into the film more quickly. This naturally results in more rapid toning, and quick toning does not yield as good prints as a slower and more gradual building up of the color image.

Having the print ready for toning as here outlined, take 1 dram of the uranium solution, add ½ dram of acetic acid and then 5 ounces of water. In a separate graduate put 1 dram of ferricyanide solution and 5 ounces of water. Just before toning, pour these two solutions together into the third graduate and use immediately. To proceed, lay the rewetted print face up in a clean tray and flow the freshly made toning bath (the two solutions combined) over the print. The print and tray must be kept in motion by gentle rocking during the toning operation. The toning solution tends to throw a red precipitate as it works. This precipitate should not be permitted to settle on the face of the print. Some workers tone their prints face down, but I do not advocate this, as it is important to take the print from the toning bath at just the right moment, and, as the toning process is short (six or seven minutes is usually sufficient even for the deepest red) you need to watch the print all the time. In the toning operation note that a constant quiet motion of the tray, to keep the solution moving over the print, is essential to success.

I have already given, in an earlier paragraph, the order in which the colors come. But that order was for a normal print. Some prints behave differently, and it is in the control of these unavoidable variations with different prints that skill and success come. A print of a half-tone subject against a jet-black background, a portrait, for instance, will hardly follow the normal order in the appearance of colors. This because the half-tones will be brown and even red-brown before the toning solution has changed the dense black deposit of the background at all. If the toning was stopped at this stage, some very pretty effects in double toning might result.

From this explanation of the toning process, the discerning reader will perceive the need for caution in selecting the best kind of a print for uranium toning. Thus a print which has a bald-headed sky will tone only in the body of the print, but if there is any tint at all to the sky, it also will tone, giving an effect not much to be desired except for sunset or sunrise pictures. If white high-lights are desired in the toned print, they must be white originally and not the least bit fogged. As double-toned effects in a print are not usually desirable, those prints having deep black shadows or dark masses will be avoided. The best kind of print for this method of toning is one fully exposed and slightly under-developed, since, when the uranium does take hold of the shadows, it makes for an increase of contrast.

Experience is the best teacher, and I could not begin to describe in detail what the reader can himself ascertain from a few experiments. Some prints needing contrast should be carried far in the toning solution; others, not needing contrast, will give better results if they are toned only through the browns, and so on. The reader who can spend a Saturday afternoon with a few bromide prints, varying in character, will learn more from his experimenting than I could tell him in many pages. For these experiments waste or imperfect black prints can be used with practical economy, the chief object being to watch the progress of toning and chemical changes.

When the desired tone is reached, remove the print from the toning solution and wash quickly and well in running water for fifteen minutes. If washed too long, the color of the print will fade and a dead and lifeless print will result. If not washed long enough, the yellow of the ferricyanide will remain in the print, robbing its gradations of brightness and purity of color and impairing the permanency of the print.

A big advantage in this method of toning is its wonderful adaptability. There is no hard and fast rule as to the proportion of the chemicals to the bulk of water used. Try two drams of each of the two solutions; then three drams of each, but watch that the print does not get beyond you in toning. The only practical difference in my formula and others that I have seen is that I make my stock solution weaker than that ordinarily advised and use less of it to a certain amount of water, because I prefer slow toning and the accompanying ease of control which the flash-in-the-pan formula does not admit. Quick toning, like quick development, tends to block the shadows in the print, and if you once get bronzed shadows the print is practically hopeless. Not quite ruined, however, as a bath in a 5 per cent solution of sodium carbonate will discharge the color and then, if the print is faded, it may be redeveloped in an alkaline developer such as metol-hydro. But before it is retoned the print must be thoroughly washed, as the presence of sodium carbonate does not permit the toning solution to do its work.

Finally, I may say that, while a bath of acetic acid and water is often advised to stop the toning action in this method, I have never found it necessary.

All the thin varieties of bromide paper curl badly in drying. If they are to be kept unmounted it is well to immerse them in water to which has been added a few drops of glycerine. This will ensure their lying flat after drying. A solution of 2 ounces of glycerine in 25 ounces of water is advised when it is desired to make bromides on heavy rough paper remain flat, after drying, for book illustration and similar purposes.

If one is trying to rush through a bromide print, it can be trimmed while wet by placing it on a sheet of stiff paper and cutting through both.

The paper will be found to cockle the mounts badly in drying. Aside from the glue mountant, formula for which accompanies the paper, I know no preventive except to mount the prints while dry with the dry mounting tissue. As the paper when wet stretches one way considerably, as much as a third of an inch on a ten- or twelve-inch length, provision must be made in trimming, especially if mounts with centers of a given size are used.

The paper being covered with an emulsion which in warm weather is very soft while wet, mounting is somewhat more difficult than with some of the other papers. My method is to mount not more than half a dozen at once, placing them face down, one on top of the other, on a glass or ferrotype plate, blotting off the surface water and spreading the paste over the top one in the usual way. I place this on the mount and then stretch over it smoothly a damp handkerchief or piece of very thin rubber cloth, rubbing the print down with my hands, seldom using the squeegee and then very lightly. By this method abrasion of the surface seldom results and air-bells are unknown. Owing to the strong contracting power of the paper in drying, the mounting paste must be used freely, especially at the edges of the print.

Apart from the methods of procedure here given, there are innumerable modifications covering every detail of contact printing and enlarging on bromide paper. Most of these have been given careful trial as published, but few have quite fulfilled the expectations they created.

BOOKS.

Manuals dealing with the manipulation of the various brands of paper may also be obtained, generally *gratis*, from the various manufacturers of bromide paper or their American agents as follows: The Eastman Kodak Co., Rochester, N. Y.; The Defender Photo Supply Co., Rochester, N. Y.; J. L. Lewis, New York City.

 # ROYAL ALBUMS

combine all the good features you seek in an album for the proper display and preservation of your

UNMOUNTED PHOTOGRAPHS AND POST CARDS

Quality in Materials
The Best of Workmanship
Pleasing in Design and Appearance
Durability in Service
Moderate in Price
A Wide Variety in Styles and Sizes
Bindings in All Desirable Colors and Finishes
Leaves and Covers of Harmonious Coloring

Ask Your Dealer to Show You
ROYAL ALBUMS

MANUFACTURED BY
F. L. SCHAFUSS COMPANY
200 FIFTH AVENUE, NEW YORK
Makers of Albums Exclusively

TO PROFESSIONAL AND AMATEUR PHOTOGRAPHERS

Write us at once for particulars regarding our
new Multi-Speed Flashlight Attachment for our

MULTI - SPEED SHUTTER

It is a simple device operating in conjunction with the Multi-Speed Shutter and attached to same with a screw on the back of the shutter.

It makes high speed photography practicable under all conditions. Takes photographs of moving objects in 1-1000th second.

Especially valuable for making instantaneous full-timed pictures of babies, aged and nervous people, groups, etc. Invaluable to press photographers for making race starts at night, indoor athletic events and all kinds of newspaper illustrations hitherto impossible.

Send for a free Specimen Photograph and Detailed Information

THE MULTI-SPEED SHUTTER JUNIOR

Simple, Efficient, Moderate in Price

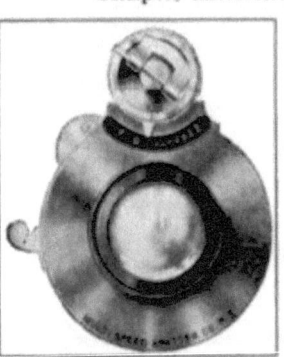

Based on the principle of the famous Multi-Speed Shutter.

Gives accurate exposures up to 1-500th second, in addition to the usual slower speeds.

Extremely compact for the small camera. Easy in operation.

Highest efficiency in illumination and definition.

Widely endorsed as the simplest and best shutter sold even at a higher price.

Send for Detailed Information and FREE Illustrated Catalogue

MULTI-SPEED SHUTTER CO., 317-323 East 34th Street, NEW YORK

LANDE'S
Patent Enlarging Lamp
PATENTED

Puts Enlarged Pictures Within Reach of Everyone. Converts any Camera into an Enlarging Lantern without the use of Condensers.

Showing Camera and Enlarger in position

Showing Back Open and Interior Arrangement

Enlargements made at home in any room which can be darkened. By means of this lamp and his regular camera any amateur is equipped to make enlargements on bromide paper. It is fitted with two incandescent gas lamps. The light is reflected so as to give a perfect even illumination over the entire surface of the negative.

The time of exposure varies according to the power of the light, the density of the negative and size of the enlargement. Approximately an average negative enlarged to four times the original size will take about one to one and one-half minutes.

Prices

No. 1. For 5 x 7 and smaller negatives $12.00
No. 2. For 8 x 10 and smaller negatives.................... 15.00

Ask your dealer to show you

MANUFACTURED BY

GEORGE MURPHY, Inc., 57 E. Ninth St., New York

Cheap Blotting Papers are expensive in use and may ruin your prints. There is practical economy and comfort in the use of

PHOTO FINISH WORLD BLOTTING

The Best for Twenty Years

Made expressly for photographic use. 100% pure cotton fibre, very absorbtive and non-linting. Absolutely free from anything that can injure the photographic print. Notable for its smooth, hard finish, giving durability in use. Try it and you'll use no other.

UNEQUALLED IN QUALITY
ECONOMY and SATISFACTION

Your dealer can supply you if you will insist on getting "Photo Finish World." It is worth insisting upon.

Manufactured by
THE ALBEMARLE PAPER MFG. CO.
Richmond, Va., U.S.A.

Produces essentially the same results as other **coal-tar** developers. It is economic by reason of slow exhaustion, and its solutions are stable for many months, the bottles need not be kept filled to the neck.

It produces fog-free negatives with the right amount of density in the high lights, a wealth of detail in the shadows and great softness.

BEST FORMULA FOR PLATES, FILMS AND SLOW-WORKING, CONTRASTY GASLIGHT PAPERS:

Hot Water	32 oz.
Duratol	15 grains
Soda Sulphite dry	¾ oz.
Potassium Carbonate dry	1 oz.
Hydroquinone	75 grains

Development complete in from 2 to 5 minutes.

AN EXCELLENT FORMULA FOR SOFT-WORKING PAPERS (ARTURO, CYKO AND ARTEX):

Hot Water	35 oz.
Duratol	15 grains
Soda Sulphite dry	1-2 oz.
Potassium Carbonate dry	¾ oz.
Hydroquinone	60 grains

Development complete in about 2 minutes.

Sample upon request from

SCHERING & GLATZ
150 Maiden Lane - - New York

Make Your Own Enlargements

The GRAPHIC
Enlarging and Reducing Camera

With this camera you can make enlargements of any size with either daylight or artificial light. No condensing lenses are required. The reflecting cone supplied with each camera distributes the light evenly over the entire surface of the negative.

PRICE

Including Negative Carrier and Kits and three sheets of diffusing Ground Glass.

No. 1.	For Negatives	5 x 7 and smaller			$28.00
No. 2.	"	"	8 x 10	" "	35.00
No. 3.	"	"	11 x 14	" "	45.00

FOLMER & SCHWING DIVISION
Eastman Kodak Co.
ROCHESTER, NEW YORK

WE RECOMMEND
"Agfa" AMIDOL
FOR DEVELOPING
BROMIDE WORK

¶ Send an "Agfa" Label and 10 cents for a copy of the "Agfa" Formulæ Book. It contains complete formulæ for all makes of paper and other valuable information.

BERLIN ANILINE WORKS
215 WATER STREET
NEW YORK

From a Small Negative You Can Print a Picture Any Size

The Radion Enlarging Printer is a device which, used in connection with your camera, produces prints from either film or glass negatives in any dimensions you wish.

The printing is almost as quick as contact printing, and the size of the prints can be varied instantly.

THE RADION ENLARGING PRINTER

enables you to get all the results of a big camera with a small and inexpensive equipment. It comes to you complete and ready to operate at any time of day or night. It is equipped with new, specially constructed Mazda lamps ready to attach to any electric light socket.

The Radion Enlarging Printer works without the use of condensers, which thereby reduces the cost of the machine and secures a better detail on the print.

PRICE $15.00

Most dealers in photographic supplies can sell you a Radion Enlarging Printer. If yours cannot, write direct to us. Write anyway for descriptive circulars that will be of great interest to every camera user.

H. C. WHITE CO.
Lens Grinders and Makers of Optical Instruments for over Forty Years

503 RIVER ST., NORTH BENNINGTON, VERMONT

Branches:
45 W. 34th St., New York City   San Francisco   London

The First Essential in Enlarging is the
Curry Utility Enlarging and Copying Board
and Springs

It is 20 x 30 inches, made of well-seasoned wood, nicely finished.

Will hold without injury any size copy of any thickness.

Utility Springs are made of the best nickeled steel.

Price, including 24 Springs, **$3.50.**

THE CURRY STEREON
Anastigmat Lens
Series A—F 4.5
Series B—F 6.3

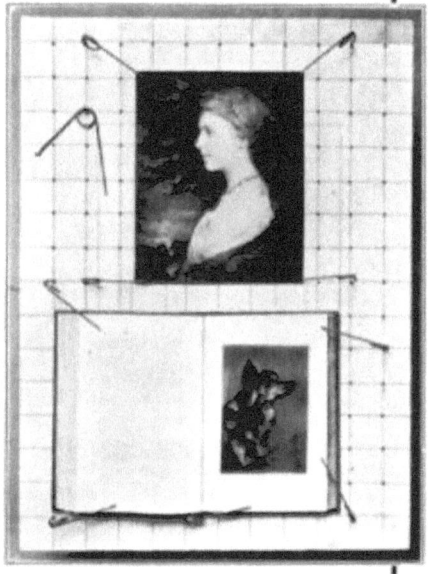

The finest lens at any price for portraits, views or enlarging.

Send for Price List and "Photo Developments."

FRANK J. CURRY

Importer, Manufacturer and Dealer in Photo Goods of Every Description.

812 Chestnut Street
Philadelphia, Pa.

For Enlarging
Bausch & Lomb-Zeiss Tessar IIb-F:6.3

because of its excellent optical corrections.

The **Tessar** has a flat field allowing exposures at full aperture.

No danger from vibration because of long exposure.

Saves exposure time on dense negatives.

For Sale by Dealers.

Send for Booklet 70 H, "Useful Tables for the Photographer," free on request.

Contains tables on enlarging.

Our name, backed by over half a century of experience, is on all our products—lenses, microscopes, field glasses, projection apparatus, engineering and other scientific instruments.

Bausch & Lomb Optical Co.
NEW YORK WASHINGTON CHICAGO SAN FRANCISCO
LONDON ROCHESTER, N.Y. FRANKFORT

You see the picture before you on the ground glass, right up to the moment of exposure and are sure of success if you use

REFLEX CAMERAS

FOR YOUR CAMERA WORK OUT-OF-DOORS AND AT HOME

REFLEX CAMERAS EXCEL
all other similar types in curtain velocity, ease and speed of operation, simplicity of construction, freedom from outside mechanism, perfect workmanship.

REGULAR
4x5 and 5x7
LONG-FOCUS
4x5 and 5x7
JUNIOR REFLEX
$3\frac{1}{4} \times 4\frac{1}{4}$ (fixed focus)
Price $12, with either one Holder or Adapter.

Ask any independent dealer or write for catalogue and sample print to

REFLEX CAMERA CO.
Newark, N. J.

GOERZ CAMERAS
ARE UNIQUE IN THE HAND CAMERA LINE

OUR NEW VEST-POCKET TENAX

is the finest pocket camera ever made. It measures, closed, $\frac{3}{4}$x2$\frac{1}{4}$x3$\frac{1}{2}$ inches. Equipped with Compound Shutter, all speeds up to 1-250 second. We furnish with it a purse case containing six nickel holders for plates 1$\frac{3}{4}$ x 2 5-16 inches. Our fine enlarging apparatus, made especially for use with this camera, will afford good enlargements of various sizes up to 5 x 7 inches.

Our **Ango, Manufoc Tenax** and **Folding Reflex** are all splendid cameras for the professional photographer and ambitious amateur.

They show the same high qualities of workmanship which distinguish

Get our new illustrated catalog from your dealer or direct from

GOERZ LENSES

C. P. GOERZ AMERICAN OPTICAL CO.
OFFICE AND FACTORY: 317 EAST 34TH STREET, NEW YORK

Dealers' Distributing Agencies: For Middle West, Burke & James, Chicago. Pacific Coast, Hirsch & Kaiser, San Francisco.

HOW TO MAKE YOUR CAMERA PAY

Here are books you'll surely want to read! For they show how you can increase your ability to make money with your camera. They tell how to make prints for book, magazine and newspaper publishers, and show how you can make money during spare time.

The LIBRARY of AMATEUR PHOTOGRAPHY

contains the boiled-down experience of years. Collier's, Leslie's and dozens of publications are always glad to consider any interesting photographs. They pay from three to five dollars apiece. A score of other profitable branches are covered. Here are the money-making facts you have wanted. These books are just from the press.

Now is the time for you to get full information. We have interesting literature to mail upon receipt of your postal. Write now for our special offer to LIVE AMATEURS.

AMERICAN PHOTO TEXTBOOK CO.
608 Adams Avenue   SCRANTON, PA.

Enlarging Simplified

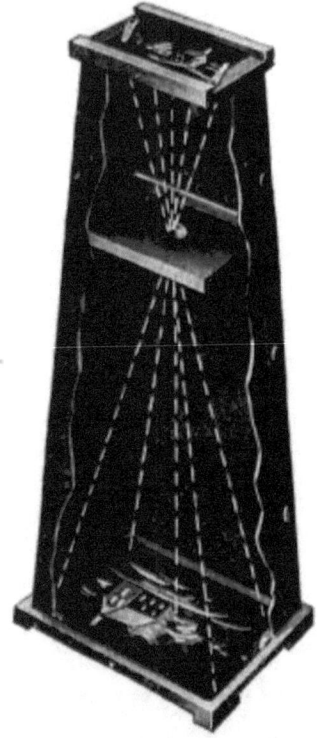

The all by daylight way of making enlargements is with the

Brownie Enlarging Camera

No focusing—no dark-room—no experience necessary when Velox is used. Just the few simple directions for finishing the prints.

The enlargment retains all the quality of the negative. The results are certain.

THE PRICE

No. 2 Brownie Enlarging Camera, for 5 x 7 Enlargements from 2¼ x 3¼ negatives, ... $2.00
No. 3, ditto, for 6½ x 8½ Enlargements from 3¼ x 4¼ negatives, ... 3.00
No. 4, ditto, for 8 x 10 Enlargements from 4 x 5 negatives (will also take 3¼ x 5½ negatives), ... 4.00

EASTMAN KODAK COMPANY,
ROCHESTER, N. Y.

All Dealers.

Prints by Gaslight

VELOX increases the number of negatives that are worth printing from.

The results from contrasty negatives are softened by the use of "Special" VELOX—improved brilliancy is secured from flat negatives by using "Regular" VELOX.

ASK FOR THE VELOX BOOK.

NEPERA DIVISION,
EASTMAN KODAK COMPANY,
ROCHESTER, N. Y.

All Dealers.

www.ingramcontent.com/pod-product-compliance
Lightning Source LLC
Chambersburg PA
CBHW030502220526
45464CB00006B/2615